I0815380

OCEAN MAMMALS

MANATEES

by Julie Murray

Cody Koala

An Imprint of Pop!
popbooksonline.com

Hello! My name is Cody Koala

This book is filled with videos, puzzles, games, and more! Scan the QR codes* while you read, or visit the website below to make this book pop.

popbooksonline.com/manatee

*Scanning QR codes requires a web-enabled smart device with a QR code reader app and a camera.

abdobooks.com

Published by Pop!, a division of ABDO, PO Box 398166, Minneapolis, Minnesota 55439.

Printed in the United States of America, North Mankato, Minnesota.

042025
082025

Cover Photo: Getty Images
Interior Photos: Adobe Stock Photos; Shutterstock Images; Getty Images
Editors: Elizabeth Andrews and Grace Hansen
Series Designer: Candice Keimig

Library of Congress Control Number: 2024948406

Publisher's Cataloging-in-Publication Data
Names: Murray, Julie, author.
Title: Manatees / by Julie Murray
Description: Minneapolis, Minnesota : Pop!, 2026 | Series: Ocean mammals | Includes online resources and index
Identifiers: ISBN 9781098247744 (lib. bdg.) | ISBN 9781098248284 (ebook)
Subjects: LCSH: Manatees--Juvenile literature. | Sea cows--Juvenile literature. | Marine mammals--Juvenile literature. | Aquatic mammals--Juvenile literature. | Sea animals--Juvenile literature.
Classification: DDC 599.55--dc23

Table of Contents

Chapter 1

Sea Cows

Manatees are large **mammals**. They live in warm, shallow coastal waters and rivers around the world. They travel to find warmer waters in the winter. Manatees are often called sea cows.

Watch a video here!

Manatees are slow-moving animals. They only swim 3 to 5 miles per hour (4.8 to 8kph). Because of this, they cannot get out of the way of fast-moving boats. Many manatees are injured every year.

Where Manatees Live

North America

Atlantic Ocean

Africa

Pacific Ocean

South America

N W E S

Manatee Range

Chapter 2

Mighty Manatee

Manatees have round bodies. Their gray skin is thick and **leathery**. They are about 14 feet (4.3m) long. They can weigh more than 2,000 pounds (907kg).

Barnacles and **algae** often grow on manatees' bodies.
Learn more here!

Manatees have large heads and strong lips. Their lips allow them to easily grab food. Manatees have short whiskers on their upper lips. They use their whiskers to feel and sense **vibrations**.

Manatees have large nostrils. They close their nostrils when they are

underwater. They open them at the surface. They come up to breathe every three to five minutes.

nostrils
whiskers
flippers
tail

Manatees have two front flippers. These are used for **steering**, crawling, and holding food. They also have a large, flat tail that moves up and down. This pushes them forward.

When manatees are less active, they can hold their breath for 20 minutes!

Manatees **communicate** with one another. They make sounds such as chirps and squeaks. They also touch noses to "kiss" and hug with their flippers.

Chapter 3

Food

Manatees spend much of their day eating plants. They eat seagrasses, **algae**, and freshwater plants. They can eat up to 200 pounds (90kg) of food each day!

Manatees have flat teeth for chewing plants.

Chapter 4

One Calf

Females have one baby every two to five years. The calf is about 3 feet (0.9m) long and weighs 65 pounds (29kg) when it's born. It stays with its mother for two years.

Manatees can live to be 60 years old!

Complete an activity here!

Making Connections

Text-to-Self

Manatees move very slowly. If you were an animal, would you want to move slowly or would you want to move fast? Why?

Text-to-Text

Have you read any books about animals that live in the water? Were there any manatees in the books? What are some other animals that live in the water?

Text-to-World

Can you think of any animals that look like a manatee? What features do they have that are similar?

Glossary

algae – plantlike living things that are common in water.

barnacle – a small ocean animal with a shell that attaches to a surface.

communicate – to share information with others.

leathery – looking like material made from the skin of an animal.

mammal – a warm-blooded animal that produces milk for its young and is usually covered with hair.

steer – to control the direction of travel.

vibration – the quick back-and-forth movement of an object.

Index

Online Resources

popbooksonline.com

Thanks for reading this Cody Koala book!

This book is filled with videos, puzzles, games, and more! Scan the QR codes* while you read, or visit the website below to make this book pop.

popbooksonline.com/manatee

*Scanning QR codes requires a web-enabled smart device with a QR code reader app and a camera.